My Earth

By Shirley Frederick

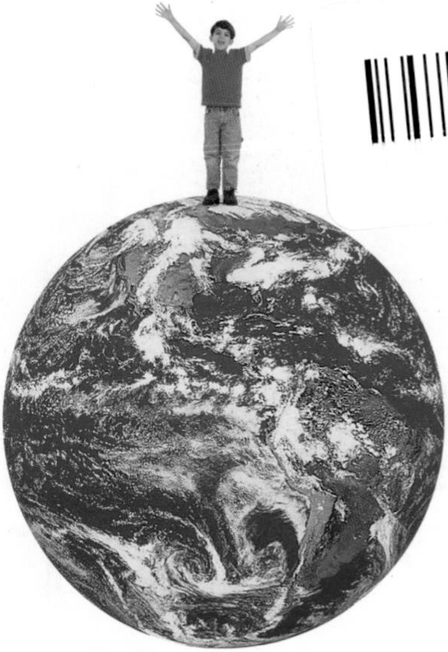

Harcourt

Orlando Boston Dallas Chicago San Diego

www.harcourtschool.com

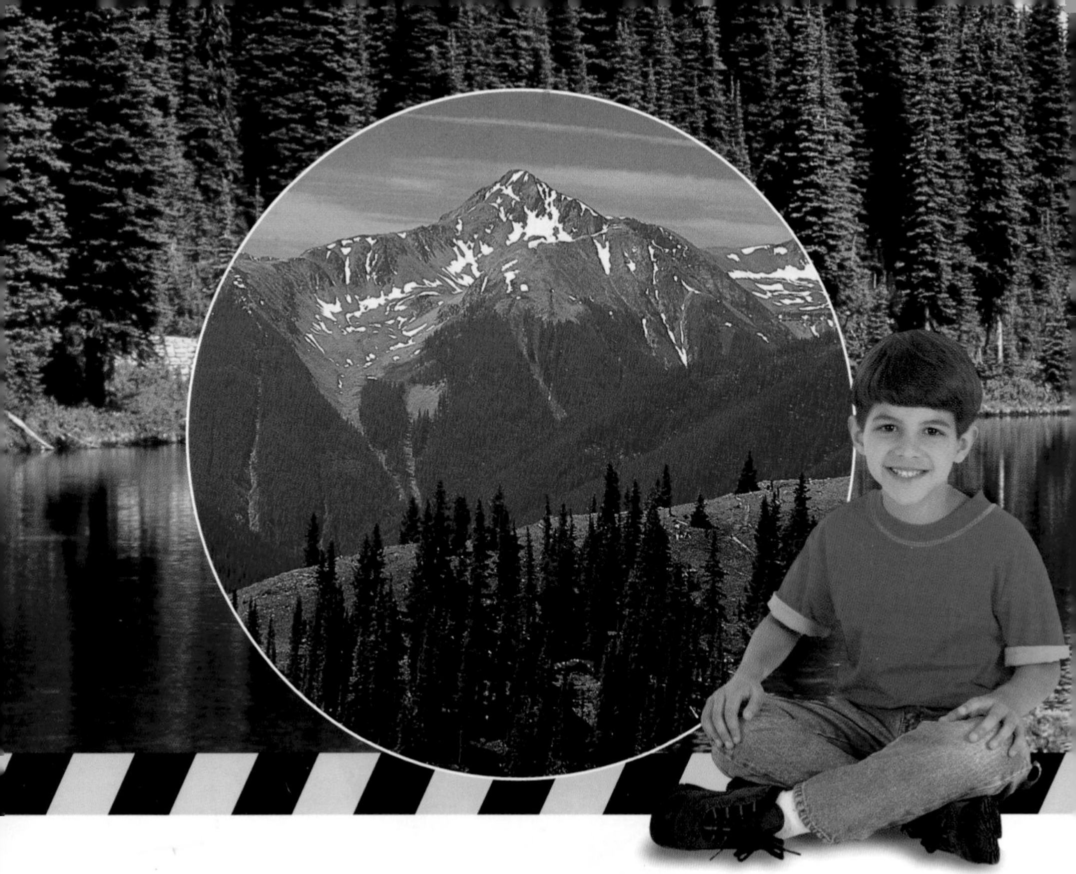

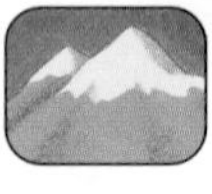

my mountain

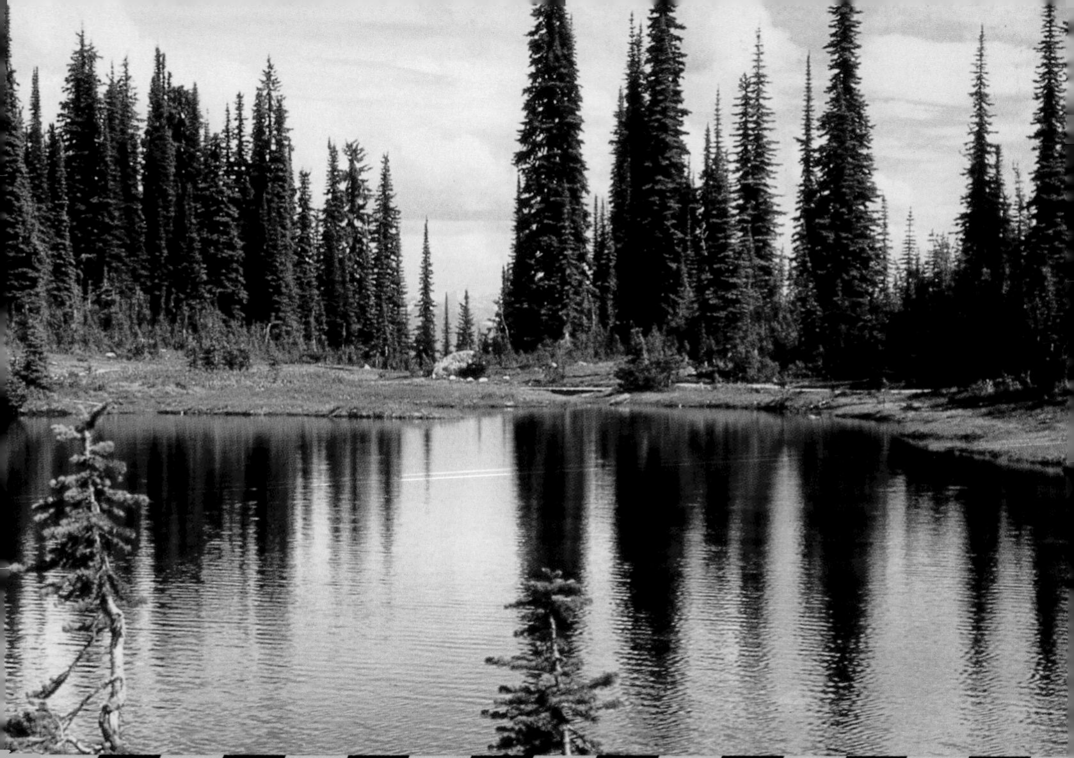

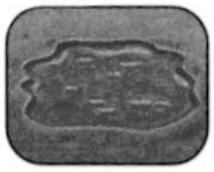

my lake

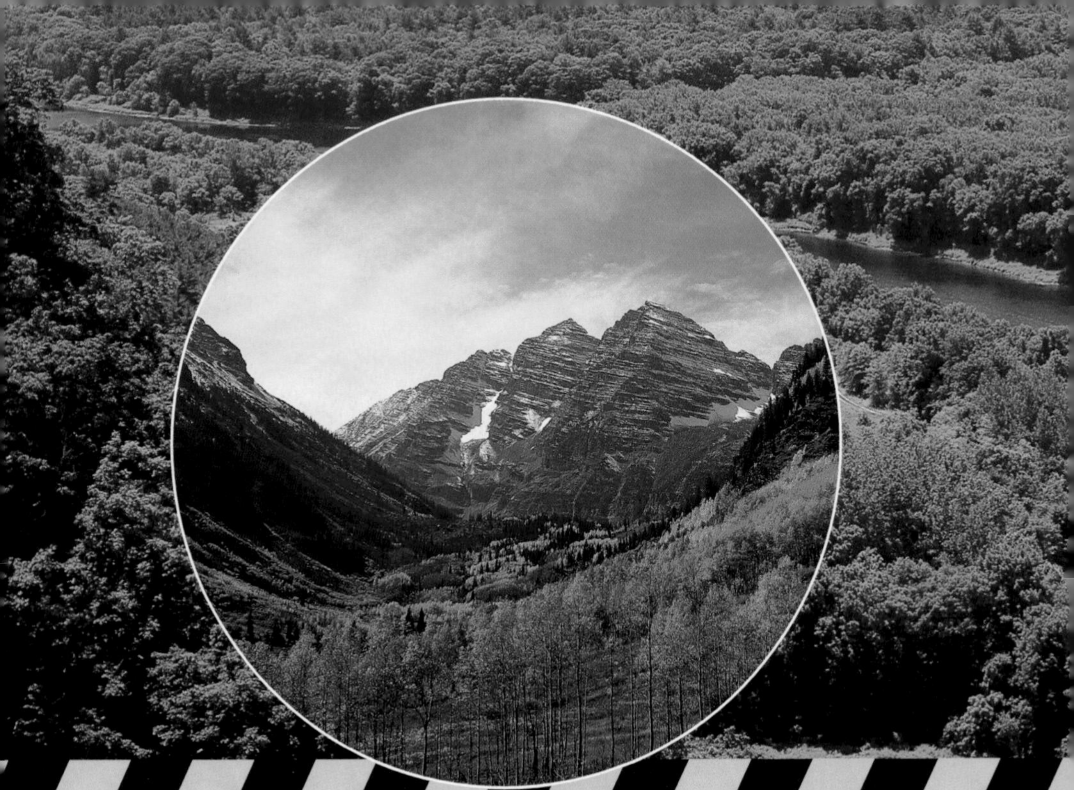

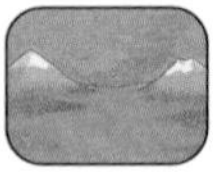

my valley

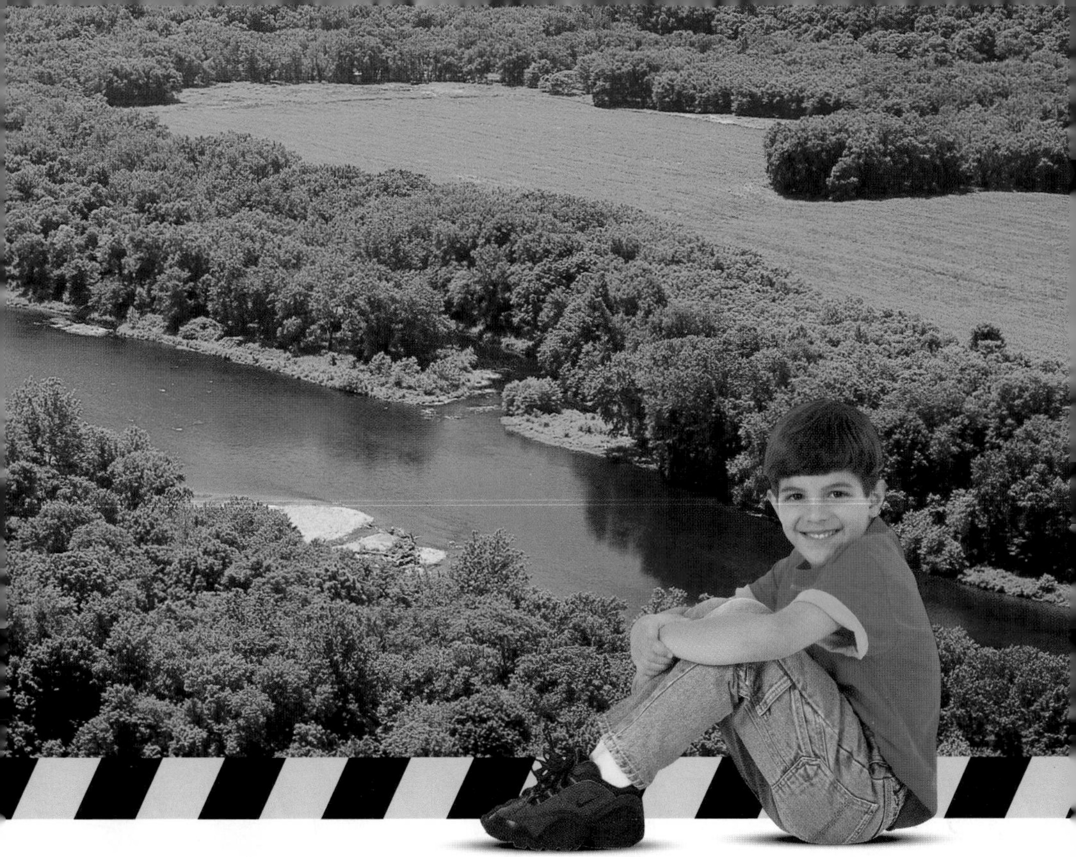

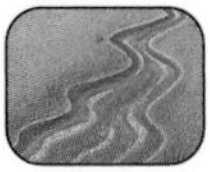

my river

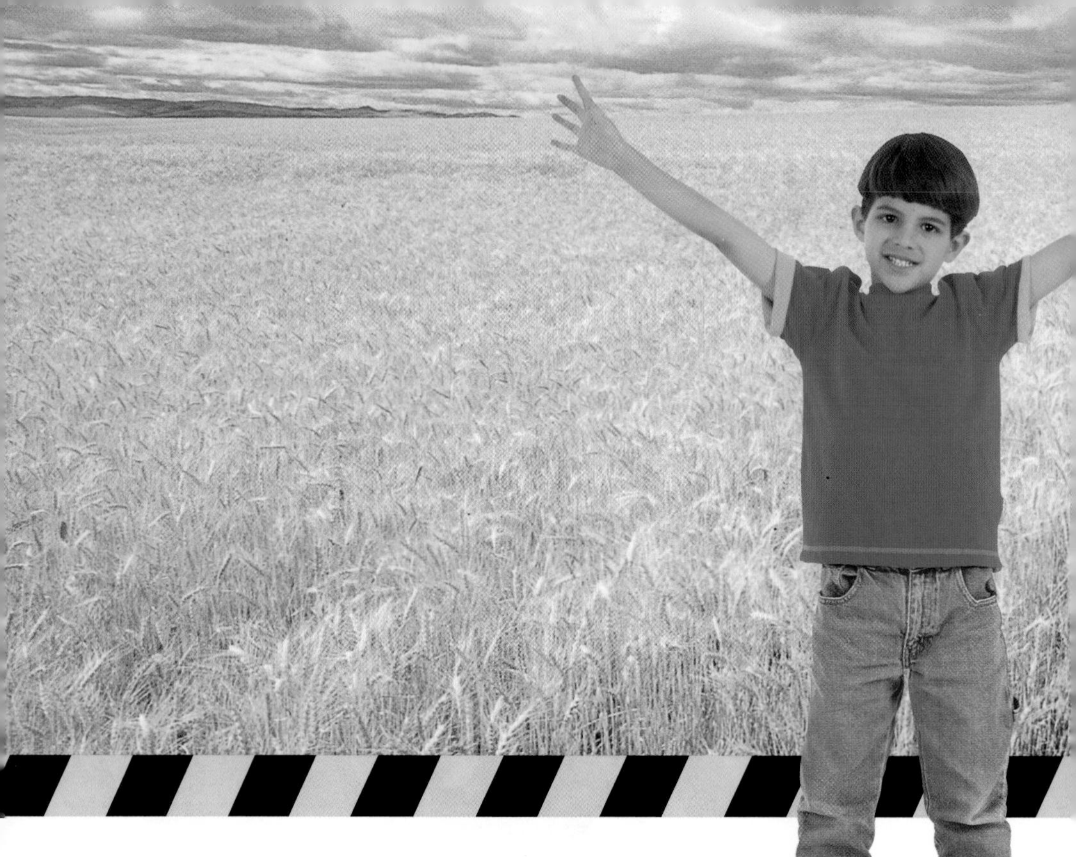

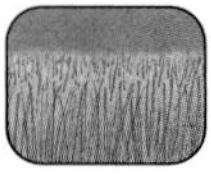

my plains

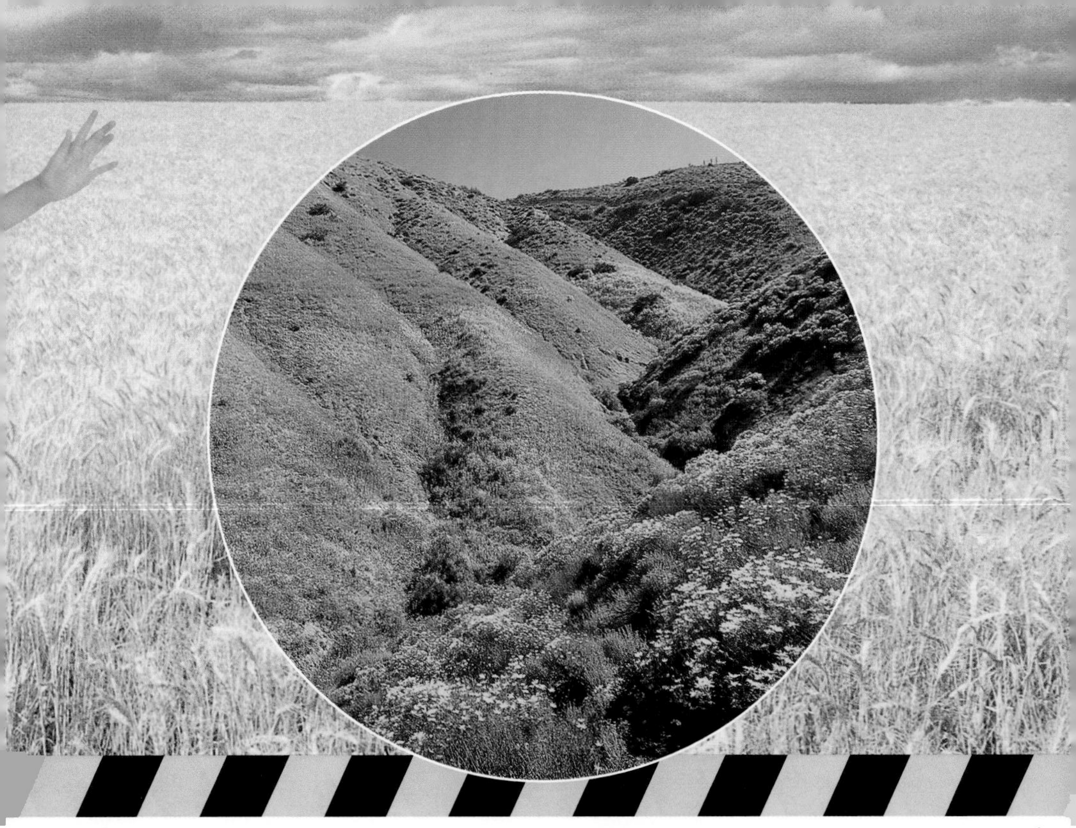

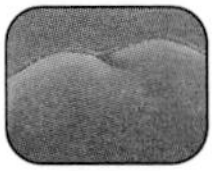

my hill

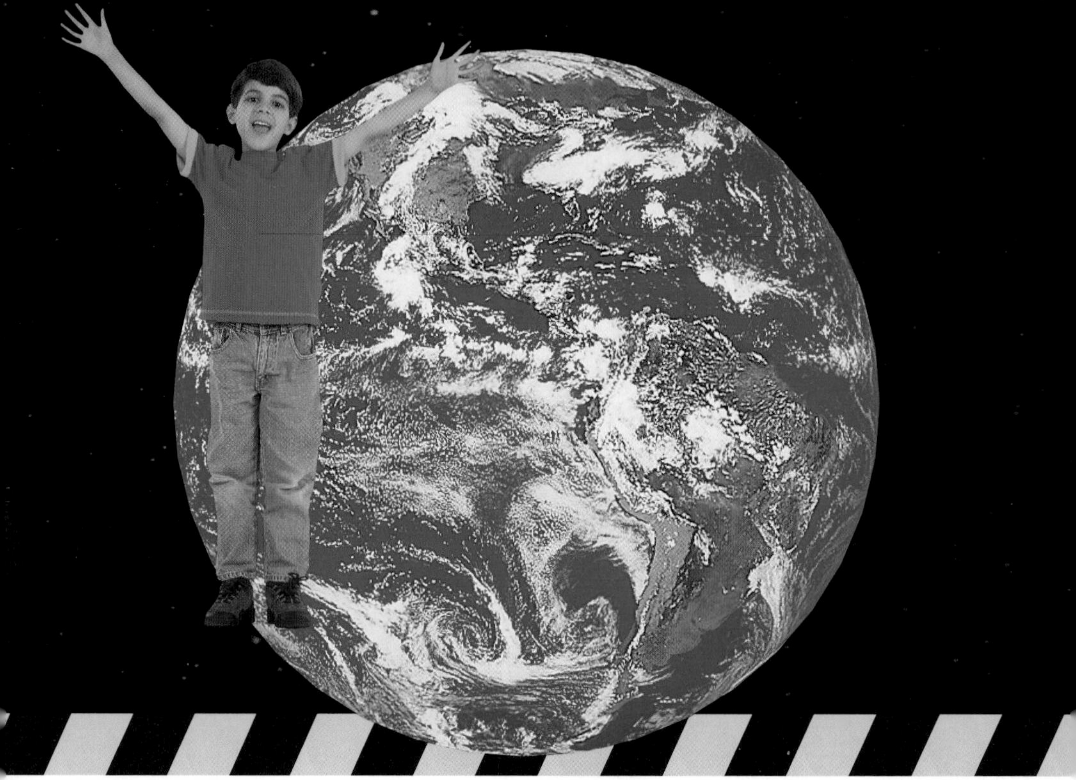

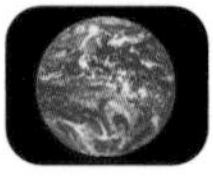

my Earth